Saved by Addition

Book One of
The Gift of Numbers
Math Fantasy Curriculum

Rachel Rogers and Joe Lineberry

Illustrations by Morgan Swofford

PROSPECTIVE PRESS ACADEMICS

PROSPECTIVE PRESS
ACADEMICS

an imprint of

PROSPECTIVE PRESS LLC

1959 Peace Haven Rd, #246, Winston-Salem, NC 27106 U.S.A.
www.prospectivepress.com

Published in the United States of America by PROSPECTIVE PRESS LLC

SAVED BY ADDITION

Library of Congress Control Number: 2016953571

ISBN 978-1-943419-35-7

Saved by Addition is the first volume in the Gift of Numbers math fantasy curriculum. For information on additional volumes in the series or for bulk sales, please send inquiries to education@prospectivepress.com

Printed in the United States of America
First softcover printing October, 2016

1 3 5 7 9 10 8 6 4 2

The text of this book is typeset in Mouse Memoirs
Accent text is typeset in Galindo

PUBLISHER'S NOTE

This book is a work of creative non-fiction with fictional fantasy elements. The people, names, characters, locations, activities, and events portrayed or implied by this book are the product of the author's imagination or are used fictitiously. Any resemblance to actual people, locations, and events is strictly coincidental. No actual numbers were harmed in the writing of this book.

Dedicated to our wonderful grandchildren:

Canaan 1

Eli 7

Emsleigh 3

Logan 4

Olivia 5

Parker 6

Rhaea ✓

Zoe 8

Before people lived on the earth, all
numbers lived in two countries —
Even Land and **Odd Nation**.

All even numbers lived in Even Land.
King 2 More was the ruler of Even Land.

2
2

3
3

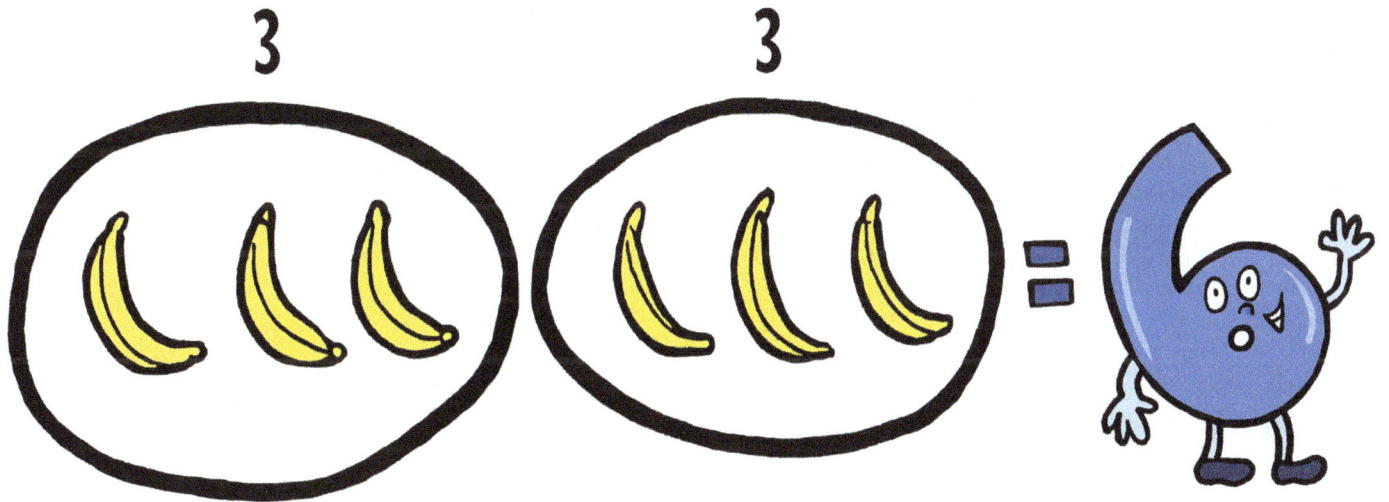

In Even Land, each number could be
divided into two equal groups
with no leftover.

All the other numbers lived across the river in Odd Nation. Odd numbers like 1,3,5,7, and 9 made their homes there. **King 1 Less** was the ruler of Odd Nation.

You know, each odd number could not be divided into two equal groups. You always have a leftover when dividing an odd number.

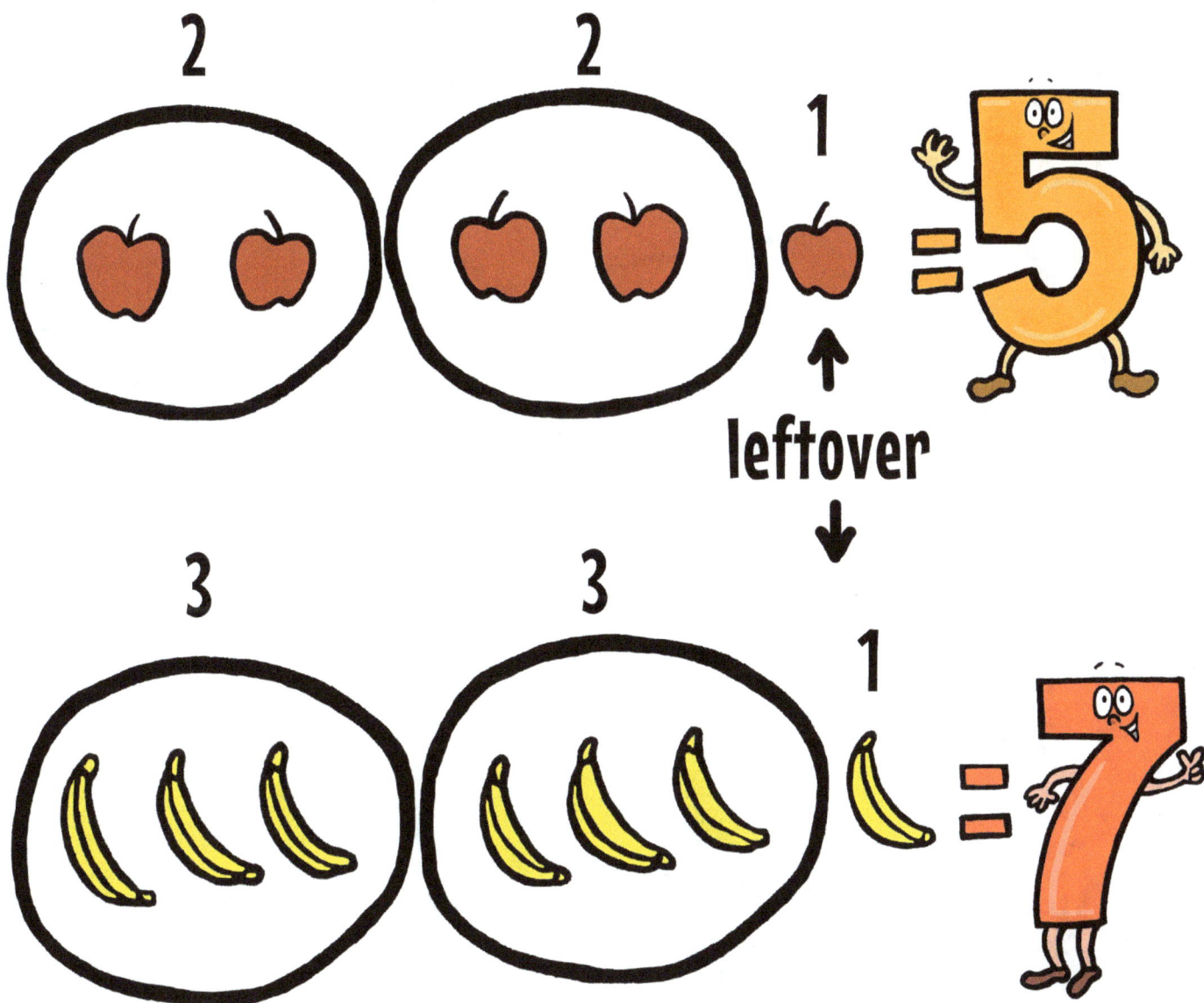

2
2
1

leftover

3
3
1

The even and odd numbers stayed in their own countries. Even numbers only wanted to play with other even numbers.

They said, "Odd numbers are so different. You cannot divide an odd number into two equal groups."

Odd numbers thought even numbers were different too. They each had separate sections at **More Hospital** on **Less Island**.

(I wonder where the hospital and the island got their names.)

One night King More had a dream. He dreamed
about a world of people, both girls and boys.

8

The girls and boys were using numbers to make life easier. They were buying toys with their money. They were also keeping score in their games. Brothers and sisters were counting the minutes before bedtime.

BALL 3 STRIKE 2 OUT 1
GUEST 0 1 0 0 2
HOME 2 0 0 0 3

But they had a big problem.
They did not have enough
numbers. Girls and boys
were sobbing. They did
not know what to do!

King More woke up with a splendid idea. He must make more numbers! He was not sure how, because no one had ever made new numbers before.

He shared his dream with **Doctor 8 Even** at More Hospital. After a few days in the lab, Doctor Even exclaimed, "Let's make a new operation! It will make More Hospital famous. We will call it 'addition.'"

Let's pause our story a moment to talk about operations. As you know, people doctors do surgical operations. Well, number doctors do math operations. And this was the first one.

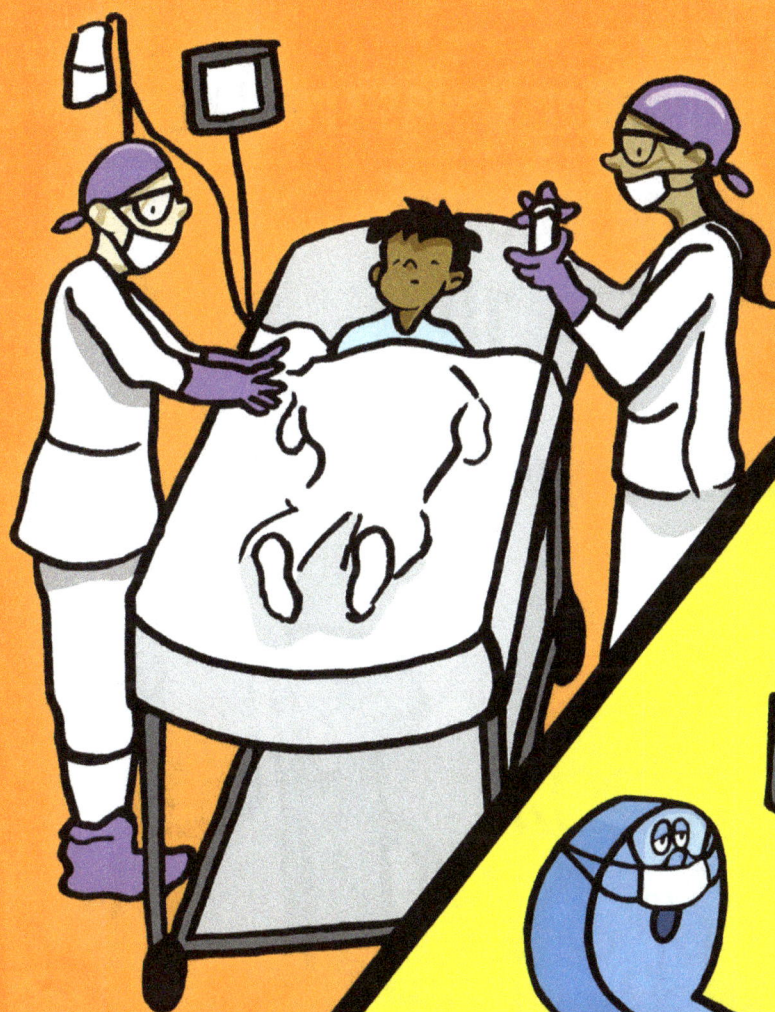

Back to our story.
Doctor Even invented a new math operation.
It was called "addition." Numbers 2 and 6
volunteered for the operation.

Doctor Even added:

$$2 + 6 = 8$$

So a new even number 8 was created.

Then he added:

$$8 + 6 = 14$$

Thus, he made a new number 14.

Even numbers from all over the country joined the addition operation. Even Land was growing with new even number children every day.

Of course, King More changed the name of the hospital to **More Children's Hospital**.

King Less of Odd Nation saw the story.

He knew future girls and boys would
need more odd numbers too.

He called **Doctor 5 Odd** at the hospital. He asked him to try adding odd numbers to create more odd numbers. Numbers 1 and 5 volunteered. Doctor Odd added them together:

$$1 + 5 = 6$$

What a surprise!

Their addition operation created a new even number 6, not an odd number.

Doctor Odd was not going to give up easily.
He tried again with other odd numbers:

$$7 + 3 = 10$$

And then he tried again:

$$5 + 9 = 14$$

Wow, he was frustrated! More and more even numbers were created with each operation.

Of course, the new numbers 6, 10, and 14 moved to Even Land. The addition operation did not add more odd numbers to Odd Nation.

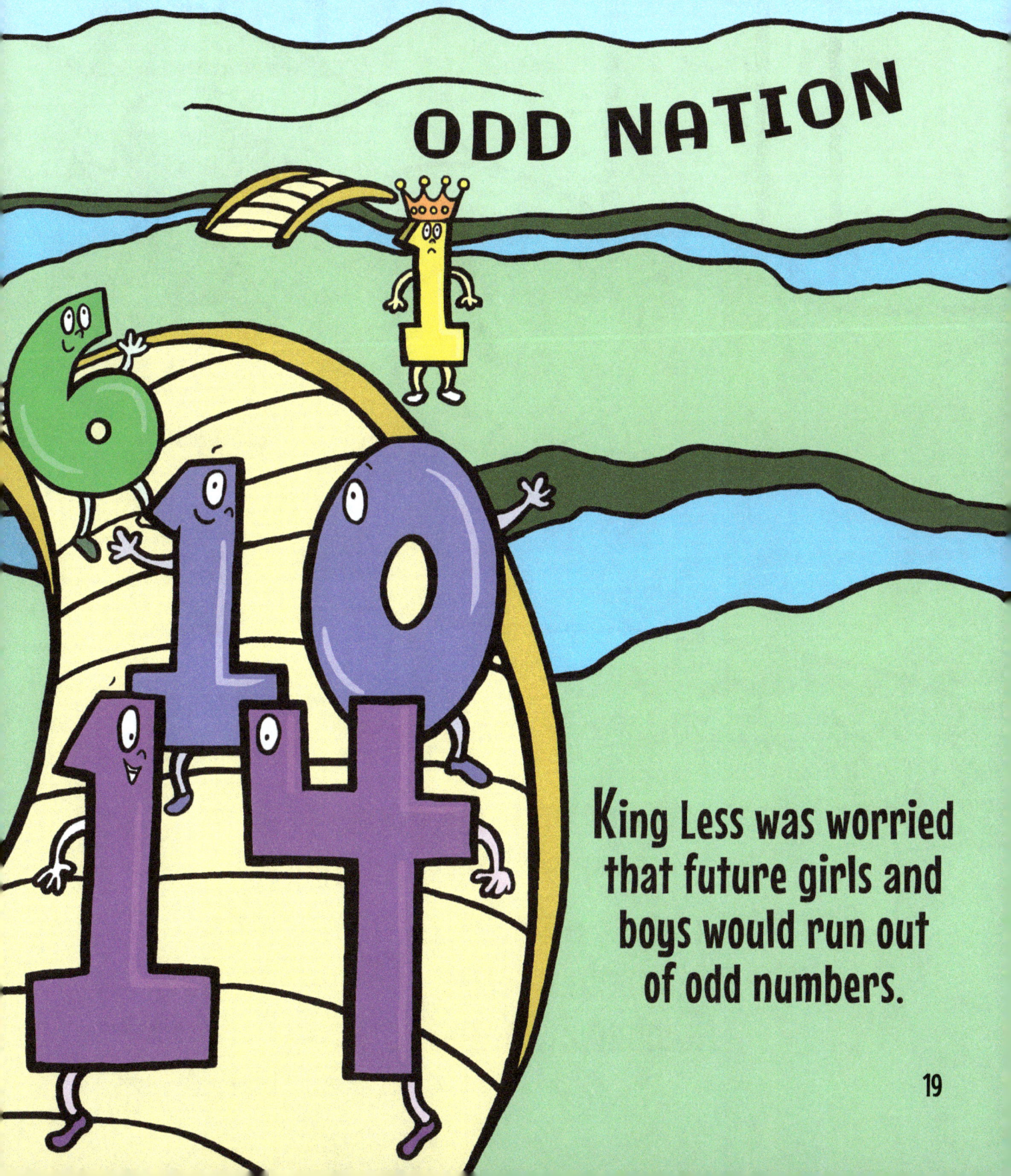

ODD NATION

King Less was worried that future girls and boys would run out of odd numbers.

Some things just work out for good.
Here is what happened. Young number
6 and his friend, number 2, sneaked into
the odd number section of the hospital.
Number 6 wanted to visit his creator,
Doctor Odd.

To avoid being caught among the odd numbers, number 6 bounced around on his head, looking like number 9. Number 2 flipped over and looked similar to number 7.

When they bounced down the odd number hall,
they bumped into Doctor Odd and King Less.
Number 2 hid behind a plant, but
number 6 flipped back over
to show who he was.

"You looked just like number 9," shouted Doctor Odd. "Hey! I just had a funny idea. I wonder what happens if we add 9 plus 6 together. That number could look cute."

So they tried it. King Less found number 9, who agreed this operation could be fun. So Doctor Odd added them together:

$$6 + 9 = 15$$

King Less gasped. Doctor Odd cried for joy. They had made the first new odd number! Number 2 came out of hiding, so Doctor Odd tried adding him to an odd number. It still worked:

$$2 + 7 = 9$$

King Less threw a
fun party for Doctor Odd and
number 6. They ate fresh fruit with
creamy yogurt. Everyone was so excited that
odd number children were filling up the hospital.

The addition operation made more even and odd numbers every day. Future girls and boys would have plenty of numbers to make their lives fun.

Saved By Addition Exercise

color the odd numbers from 1-49 gray

color the odd numbers from 50-99 yellow

color the even numbers from 1-49 red

color the even numbers from 50-99 blue

1	2	3	4	5	6	7	8	9	10	11	12	13	14	15	16	17	18	19	20
									22	14									
									38	8	20	18							
									44	24	36	16	10						
									51										
								81	77	75									
							67	79	99	69	65								
						73	85	83	61	71	89	87							
					63	53	55	59	57	93	91	97	95						
							70	98	66	52	70	66	94						
							88	54	60	90		82							
							64	84	54	78		74							
							52			80	66	84	94						
50		82		70			60			86	58	92	54		68		72		88
64	98	66	52	42			90	76	50	54	78	72	68		94	56	90	76	84
84	56	92	80	62			96	56	58	88	76	54	70		60	78	70	54	50
	66	88	90				56	52	78	92	82	66	50			66	86	72	
	98		78	49		33		21		15		5		47		23	90		80
	62		54	3	17	11	35	7	33	11	27	45	47	7	17	49	54		64
	52		74	27	13	41	37	9	27	37	21	39	19	13	27	9	58		60
	80	96	72	33				21	3	17	35	23				25	50	62	70
	76	60	50	45				33	26	12	32	17				31	64	74	78
	68	94	66	39	15	39	9	37	4	42	28	19	25	7	19	47	88	82	96
	96	78	74	41	11	1	41	49	40	46	30	5	43	1	5	21	86	90	56
	58	56	90	23	27	17	23	43	6	48	12	25	39	37	3	33	58	52	64
	92	62	84	31	29	3	29	45	34	22	2	41	11	47	15	45	82	60	98

Discussion Questions

1. How can you tell that a number is even? Odd?
2. How did the even and odd numbers change from the beginning to the end of the story?
3. Why is *Saved by Addition* a good title for this book?
4. Why do you think the authors wrote this book?

Saved by Addition Solution

About the Authors

Rachel Rogers
is a 2nd grade teacher at Old Richmond Elementary School, Winston-Salem, NC. She has more than 35 years of experience teaching first, second, and third graders.

Joe Lineberry
told similar stories to his sons when they were growing up. He is also the author of *Let's Stop Playing Games: Finding Freedom in Authentic Living.*

www.ingramcontent.com/pod-product-compliance
Lightning Source LLC
Chambersburg PA
CBHW051802200326
41597CB00025B/4656